Against the Clock
Passage of Time

JD Arden

Preface

Time is the one thing we are given in equal measure, yet how we perceive and use it differs drastically from one person to the next. It's a relentless force, ticking away quietly in the background, shaping our lives and decisions without us even realizing it. Some people treat time as an enemy, constantly racing against it, while others see it as a companion, savoring each moment. Yet, no matter how we view it, time remains undefeated.

In "Against the Clock", I invite you to join me on a philosophical journey that examines time's profound impact on our existence. From the way we measure it to the way it influences our choices, I explore how our relationship with time defines who we are. This book isn't just about the passing minutes or hours. It's about understanding how the very concept of time shapes our deepest fears, our most cherished memories, and our wildest dreams.

In these pages, we'll question whether time is truly linear or if it bends and stretches based on our perceptions and experiences. We'll confront the anxiety that comes from our awareness of time's passing, and we'll look at the moments that seem to last forever, leaving an indelible mark on our hearts and minds. This isn't a book of answers, but rather a space to reflect on the questions that time raises and the ways it shapes us.

Time may be infinite, but our lives are not. As you read on, I hope you find yourself contemplating the ticking clock in a new light, seeing not just the minutes slipping away, but the rich possibilities held within each one.

Table of Contents

Preface II

Chapter 1: The Nature of Time: A Relentless Companion 1

Chapter 2: Measuring the Immeasurable: How We Track Time 4

Chapter 3: Linear or Circular: Is Time Truly Straightforward? 7

Chapter 4: Living in the Moment: The Pursuit of Presentness 10

Chapter 5: The Fast Lane: Racing Against Time 14

Chapter 6: Time and Memory: The Past, Preserved and Faded 17

Chapter 7: The Anxiety of Time: Confronting Mortality 21

Chapter 8: Time as Currency: The Economics of Minutes and Hours 25

Chapter 9: Suspended Moments: When Time Stands Still 29

Chapter 10: The Elasticity of Time: Why Time Feels Different for Everyone 33

Chapter 11: Out of Sync: The Disruption of Time in Modern Life .. 38

Chapter 12: Borrowed Time: Living with Regret and Anticipation . 43

Chapter 13: The Gift of Time: Finding Meaning in the Moments We Have 47

Chapter 14: The Timeless Self: How We Outlive Our Own Time .. 51

Chapter 15: Beyond the Clock: Rethinking Time's Hold on Us 55

Conclusion 58

End Note 59

Chapter 1:
The Nature of Time: A Relentless Companion

Time is the one constant that touches every human life, a force both invisible and undeniable. It shapes who we are and dictates the rhythm of our existence, yet we can't fully grasp it. Time is there in the quiet moments as we lie awake, and it's there in the frantic rush of our daily lives. It's relentless, indifferent, and inescapable, defining every choice we make. And yet, if someone were to ask what time really is, we might struggle for an answer. Is it something we've created, or is it an unyielding reality that exists outside our understanding?

Physicists tell us that time is another dimension, like space, but one we can't move around in freely. It's the framework within which we experience life, giving shape to the past, the present, and the future. Einstein famously showed that time isn't fixed; it can bend and stretch depending on speed and gravity. To us, these concepts might seem abstract, something to ponder only briefly. In our everyday lives, time seems steady and straightforward, marching forward, never pausing. But time's true nature is far from simple, and as much as we try to master it, time often ends up mastering us.

For most people, time is more than clocks and calendars. It's how we feel life passing by. Sometimes, hours seem to drag, as though each minute is being stretched unbearably thin. Other times, they vanish before we even notice. Our sense of time changes as we age. Childhood feels endless, days stretching on with new discoveries and experiences. But as we grow older, time seems to speed up, days blending into months and years with a dizzying quickness. This isn't just an illusion; it's a reflection of how we live. When we're young, everything is new, and our minds are busy processing it all. As we age, routines take over, and fewer things surprise us. With routine comes the feeling that time is slipping away.

Time also molds our identity. We often think we're making choices freely, but those choices are deeply influenced by the ticking clock. We might settle into a career, not out of passion, but because time seems to be running out. We shape our lives around invisible milestones, measuring success by how closely we follow this silent schedule. Start a career by 25, find love by 30, have a family soon after. When we miss these checkpoints, it's easy to feel like we're lagging behind. But the truth is, there is no universal clock. This sense of being "on time" or "behind" is something we impose on ourselves, a trick we play to feel like we're in control.

And yet, control is precisely what time denies us. We schedule our days, trying to box our lives into neat little compartments, thinking we're masters of our own fate. But one twist of fate can shatter those illusions. Time isn't something we command. It can shatter plans, disrupt dreams, and force us to adjust. A diagnosis, a sudden loss, a chance encounter—any of these can upend the life we thought we had mapped out. We live as though we have endless tomorrows, always assuming there will be time later for what matters most. But time waits for no one, and sometimes, it leaves us behind before we realize it.

Maybe it's better to think of time as a companion, one that walks alongside us rather than something we're racing against. Time doesn't change; we do. We age, our faces carry more lines, and our experiences pile up. All the while, time continues, indifferent to our struggles or successes. People say time heals, that it can soften our pain and allow us to grow. And while this is true, time is also ruthless. It takes as much as it gives, erasing memories and altering people. Time is a constant presence, both comforting and brutal, reminding us that nothing lasts forever.

Time's relentlessness is its cruelty, but it's also its beauty. It drives us to move, to reach for something more, to grow and evolve. The fleeting nature of life makes each moment precious. It's the reason we cherish love, why we pursue dreams, and why we find joy in small, simple pleasures. Time unites us all; rich or poor, strong or weak, we all feel its pull. We are all against the clock, yet perhaps we are also with it. What if, instead of fearing time's passage, we learned to embrace it?

The relentless ticking isn't an enemy. It's a gift, a reminder that each moment is finite and worth savoring. By accepting time for what it is—a force beyond our control but within our reach to experience—we might find a way to live more fully. Time, after all, doesn't care if we fight it or if we go along with it. It just is, unwavering and indifferent, moving ever forward. The question is not whether we can conquer time, but how we will choose to live within its bounds.

Chapter 2: Measuring the Immeasurable: How We Track Time

It's a curious thing, our obsession with measuring time. We map it out with clocks and calendars, trying to tame it with numbers and divisions. Seconds, minutes, hours, days these are the tools we use to quantify something that, by its nature, defies boundaries. We take comfort in this precision, believing that if we can just keep track of time, we can somehow control it. But does slicing time into neat segments really help us, or does it just deepen the illusion?

The history of timekeeping is as old as civilization itself. Ancient cultures looked to the sky, marking the passing of days with the sun's movements and the change of seasons. They built monuments to track solstices and equinoxes, aligning their lives with the natural cycles of light and dark. Early calendars reflected a world where time was something observed, not controlled. Farmers, hunters, gatherers—they all respected time's rhythms, understanding that it was beyond their grasp.

As societies evolved, so did our methods of tracking time. Sundials gave way to water clocks, which were then replaced by mechanical clocks. With each innovation, we refined our ability to count time's passing, moving from the imprecise to the exact. By the time we reached the Industrial Revolution, the ticking of clocks had become synonymous with progress and productivity. Hours and minutes were no longer just markers of the day; they became units of labor, measuring our worth and output. Time, once an abstract concept, had been transformed into something quantifiable, a currency we could spend or waste.

But time doesn't care about our measurements. It flows at its own pace, indifferent to the numbers we assign to it. We may set our clocks and plan our schedules, but time keeps moving, unaffected by our attempts to pin it down. We measure time, but can we really understand it? We live in a world of deadlines and appointments, each one carefully timed and planned. And yet, even the most meticulous schedule can be upended in an instant. An hour of waiting can feel like an eternity, while a joyful moment seems to vanish in the blink of an eye.

Our efforts to measure time speak to a deeper need for order. In an unpredictable world, time provides a framework, a way to impose structure on chaos. It offers a sense of control, however fleeting. We live by the clock, trusting its steady rhythm to guide our days and nights. But this reliance on timekeeping can be a double-edged sword. It drives us to fill our hours with tasks and goals, leaving little room for spontaneity. In trying to measure every moment, we risk losing the sense of wonder that comes from simply letting time be.

Consider the way we use time as a benchmark for success. From an early age, we're taught to measure our lives in terms of milestones: graduate by this age, get a job by that age, start a family by another. These timelines, imposed by society and reinforced by our own expectations, can create a sense of urgency, a feeling that we're constantly running out of time. We rush through life, always looking ahead to the next deadline, the next goal. But in our hurry, we often overlook the moments that truly matter.

What if we stopped measuring time so precisely? What if we allowed ourselves to experience it more organically, without the constant pressure of the clock? Would we find more meaning in the present, more joy in the here and now? It's a difficult question, one that challenges the very fabric of modern life. We are so accustomed to measuring time that we've forgotten how to live without it. The idea of a day without a schedule, an hour without a task, seems almost unthinkable. And yet, it is precisely in these unmeasured moments that we often find the deepest connections, the richest experiences.

Timekeeping has its place, of course. It helps us coordinate, connect, and achieve. But there is also value in stepping away from the clock, in letting time flow without boundaries or constraints. In those moments, we can reconnect with the rhythms of life that are beyond measurement, the cycles that remind us of our place in a larger world. Time may be relentless, but it is also generous, offering us each day anew, each moment fresh and uncharted. To truly appreciate it, perhaps we need to let go of our need to measure it, to embrace the immeasurable nature of time itself.

Chapter 3:
Linear or Circular: Is Time Truly Straightforward?

Time, to most of us, feels like a one-way street. We experience it as a steady progression from the past to the present and then into the future, as if we're on a track that never loops back. This sense of linearity seems so obvious that we rarely question it. Our lives unfold in a sequence of moments, each one building on the last, pushing us relentlessly forward. Birth, growth, decline, death—the cycle seems clear and irreversible. But is time truly as straightforward as it seems?

Many cultures and philosophies have offered alternative views, suggesting that time might not be a straight line at all. In some traditions, time is seen as cyclical, a repeating series of events that follow a natural rhythm. This idea is reflected in the changing seasons, the cycles of the moon, and even in the way history seems to repeat itself. To those who see time as circular, life isn't a race to the finish line but a series of recurring patterns, each with its own lessons and purpose.

The ancient Greeks had two words for time: chronos and kairos. Chronos referred to chronological, sequential time—the kind we're used to tracking with clocks and calendars. But kairos was different. It spoke to the right moment, a kind of opportune time that wasn't measured by hours or minutes but by the significance of events. In the world of kairos, time was less about linear progression and more about moments that held meaning. It's as if, alongside the ticking of the clock, there's another layer of time that doesn't follow a straight path but instead emerges in those instances when everything aligns.

Some philosophies take this idea even further, suggesting that all moments might exist simultaneously. According to certain interpretations of physics, particularly in the realm of quantum theory, the distinction between past, present, and future is not as clear-cut as we perceive it to be. Time, in this view, is a sort of tapestry, with every thread—every moment—interwoven into a larger whole. If that's the case, then perhaps time is not a river that flows from one end to the other but a vast sea in which all moments exist together, each accessible depending on where you stand.

This notion of time challenges our understanding of reality. It invites us to consider that the past isn't gone and the future isn't yet to come but that both may already exist in some form, outside our reach but ever-present. When we reflect on memories, they feel as real as the moment we experienced them. They're not simply relics of something that once happened; they're pieces of us, still alive in some way, held within the fabric of time itself. Similarly, when we imagine the future, we're not merely daydreaming. We're pulling potential into our present, giving shape to what might come next. It's as if, in our minds, we're capable of bending time, seeing beyond the straight line and into a landscape of possibilities.

In contrast, our daily lives are dominated by the linear view of time. We structure our days around schedules, plan for the future, and try to leave the past behind. This approach has its benefits—it gives us a sense of control, a way to organize our lives and set goals. But it can also be limiting. By focusing solely on the forward march of time, we risk losing sight of the cycles that shape our existence. We forget that life is not just about moving forward but also about returning, reflecting, and renewing. We see this in nature, where cycles of birth, growth, and decay repeat endlessly. Trees shed their leaves in autumn, only to bloom again in spring. The tides ebb and flow, the sun rises and sets. Life is full of circles, reminding us that endings are often beginnings in disguise.

Whether time is linear, circular, or something else entirely, what matters most is how we choose to experience it. We can view it as a straight path to the end, rushing from one moment to the next, always focused on what's ahead. Or we can embrace the idea of cycles, recognizing that life is filled with patterns that return, giving us a chance to revisit, to learn, to grow. In the end, time may be both linear and circular, a blend of progression and repetition, a dance between moving forward and coming back to where we started.

We may never fully understand time's true nature, but perhaps that's part of its beauty. It remains a mystery, something just beyond our grasp, inviting us to wonder, to question, and to explore. By stepping back from our linear view, we might find that time is less a line and more a circle, a spiral, or perhaps something altogether different—an invitation to experience each moment fully, not just as a step toward what's next but as a part of a larger, ever-turning wheel.

Chapter 4:
Living in the Moment: The Pursuit of Presentness

"Live in the moment"—a phrase so simple, yet it carries a weight that feels almost impossible to hold. Being present seems like the easiest thing in the world, and yet, in practice, it's one of the hardest. Our minds are perpetually moving—always ahead, planning the next task, or lagging behind, replaying what has already passed. For all our talk about the importance of now, few of us truly know how to dwell in it.

Modern life does little to help. We're bombarded by distractions, constantly nudged by our devices and alerts that seem to demand our attention at every turn. It's a rare person who can focus entirely on one thing without feeling the pull of something else. Even when we try to be present, there's often a nagging sense that we should be somewhere else, doing something more. This drive to always be "on" has made being in the moment feel almost like a luxury, something that exists just out of reach.

But living in the moment is more than a luxury—it's a necessity. The present is the only time we truly possess. The past is already written, and the future is a story yet to unfold. The present is where life actually happens. It's where we experience, where we feel, where we connect. Yet, despite this, we spend so little time truly engaged with it. We're always looking forward, hoping for something better, or looking back, haunted by regrets. Meanwhile, life slips by, one unnoticed moment after another.

Being present doesn't come naturally to most of us. It's a skill that requires practice, a conscious decision to put down the phone, to stop checking the clock, and to let go of whatever else is demanding our attention. It's about immersing ourselves in what we're doing, whether it's something as mundane as washing the dishes or as profound as spending time with a loved one. When we focus entirely on what's in front of us, the world takes on a different texture. Colours seem brighter, sounds clearer, sensations sharper. It's as if we're seeing the world with fresh eyes, experiencing it as it truly is, without the fog of distractions.

But living in the moment isn't just about what we're doing. It's also about how we're feeling. Embracing the present means allowing ourselves to experience emotions fully, without pushing them away or trying to numb them. Too often, we're afraid to feel deeply. We try to avoid discomfort, to shield ourselves from pain. But by doing so, we also dull our ability to experience joy. To live in the moment is to accept the full range of our emotions, to let ourselves feel without judgment or fear. It's about being vulnerable, open to whatever the moment brings, and trusting that we can handle it.

This pursuit of presentness requires letting go of control. When we try to hold on too tightly, we become trapped in our own expectations, unable to appreciate what's right in front of us. True presence is about surrender—allowing ourselves to be swept up in the moment, without needing it to be anything other than what it is. It's about experiencing life without filters, without constantly trying to shape it to fit our desires.

There's a quiet beauty in this kind of surrender. When we let go of the need to control, we open ourselves to surprise, to wonder, to connection. We begin to see that life is not a series of tasks to be completed but a tapestry of experiences to be savoured. Each moment, no matter how ordinary, holds the potential for something extraordinary if we're willing to be fully there for it. It's in these moments of full engagement that we truly come alive, that we feel the pulse of life in its most raw and authentic form.

The truth is, living in the moment isn't about doing more or less—it's about being more. It's about being present, not just with others, but with ourselves. It's about noticing the details, the small things we often overlook in our rush to get somewhere else. It's about breathing deeply, feeling the ground beneath our feet, tasting our food, listening to the sounds around us. When we give ourselves to the moment, we discover a richness that goes beyond words. We find ourselves immersed in a world that is alive with colour, texture, and sensation, a world that exists only here and now.

Of course, being present doesn't mean ignoring the future or dismissing the past. Both have their place, and both can offer valuable insights. But the future and the past are only as meaningful as the present moment allows them to be. It's in the present that we make choices, that we act, that we decide who we want to become. The future is built from the moments we live now, and the past is enriched by how fully we experience each day.

In the end, the pursuit of presentness is not about escaping reality; it's about embracing it. It's about choosing to be here, in this moment, with all its imperfections and possibilities. It's about recognizing that life is happening now, not someday or somewhere else. To live in the moment is to honour the gift of time, to acknowledge that each second is precious and fleeting, and to understand that the only time we truly have is now.

Chapter 5:
The Fast Lane: Racing Against Time

Life today feels like a relentless race, one where the finish line keeps shifting further away. We rush from one task to the next, convinced that if we move quickly enough, we'll finally catch up. But catch up to what, exactly? The faster we go, the more there seems to be done, and the more we become entangled in a cycle that rarely allows for rest. This obsession with productivity drives us to maximise every second, but at what cost?

In the modern world, we equate speed with success. The quicker we are, the more productive we appear, and productivity has become a kind of badge of honour. We're constantly encouraged to do more, achieve more, and squeeze every drop of time out of the day. There's always another goal to reach, another milestone to surpass. And so, we push ourselves, constantly racing against the clock, rarely stopping to question whether any of this rushing is actually leading us somewhere meaningful.

The irony is that while technology has given us tools to save time, it often ends up making us feel as if we have less of it. We're connected around the clock, with devices that allow us to work, communicate, and stay informed at all hours. But this constant connectivity also means we never truly switch off. We're always available, always reachable, and always just one alert away from being pulled back into the whirlwind. In the pursuit of efficiency, we've made ourselves prisoners to the very tools that were meant to liberate us.

It's easy to see why we get caught up in the fast lane. Society prizes busyness, treating it as a sign of importance and success. A packed schedule suggests we're in demand, that we're achieving, that we matter. But busyness is often a hollow form of fulfilment. We can fill our days with activity and still feel a sense of emptiness. We can rush through life ticking boxes and meeting deadlines, all the while feeling like we're missing out on something vital.

What we're often missing is depth. The fast lane is all about speed, but it rarely allows for meaningful engagement. When we're racing against time, we don't have the luxury of savouring experiences, of truly immersing ourselves in what we're doing. Instead, we skim the surface, focusing on getting things done rather than on the quality of those things. In our haste, we overlook the small details, the moments of beauty and connection that give life its richness. We may be moving quickly, but we're barely touching the world around us.

Slowing down can feel almost counter-cultural in a world that glorifies speed. It's easy to fear that if we step out of the fast lane, we'll fall behind. But this fear is based on a false premise—that faster is always better. The truth is, speed doesn't necessarily lead to satisfaction. In fact, it often has the opposite effect, leaving us feeling exhausted, disconnected, and unsatisfied. There's a reason why many people report feeling burnt out despite their apparent success. The constant rush takes a toll, not just on our bodies, but on our minds and spirits.

Stepping out of the fast lane requires a conscious decision to slow down, to resist the pressure to keep up with the endless demands of modern life. It means choosing to value depth over speed, quality over quantity. It means recognising that life is not a competition, that we're not defined by how much we can accomplish or how quickly we can do it. Slowing down allows us to reclaim our time, to spend it on things that truly matter, and to experience life in a fuller, more satisfying way.

When we slow down, we give ourselves the chance to reflect, to connect, to appreciate. We create space for spontaneity, for moments that aren't planned or productive but that bring joy and meaning. Slowing down doesn't mean abandoning our ambitions or giving up on our goals. It simply means recognising that there's more to life than constant motion. It's about finding a balance, about learning that rest and reflection are as important as action and achievement.

The fast lane may seem thrilling, but it often leaves us feeling depleted. True fulfilment doesn't come from how quickly we can get from one point to the next, but from the richness of the experiences we have along the way. We don't have to race through life. We can choose to step off the treadmill, to set our own pace, to savour the journey rather than rush to the end. After all, the finish line is the same for everyone. The question is not how fast we get there, but how deeply we live along the way.

Chapter 6:
Time and Memory: The Past, Preserved and Faded

Time and memory are inextricably linked, two forces that shape our understanding of ourselves and the world around us. Memory is how we make sense of the past, how we hold onto moments that have shaped us, how we build the story of our lives. Yet, memory is not as reliable as we often believe. It shifts and changes, sometimes sharpening with time, and sometimes fading into obscurity. Our memories are more like paintings than photographs—sketches, coloured by emotion and perspective, altered by the passage of time.

The way we remember isn't static; it's fluid, shaped by our current state of mind and our experiences since. A memory that feels vivid and alive today may, in years to come, be reduced to just a fragment, a shadow of its former self. In this way, memory is as much about forgetting as it is about remembering. We lose pieces of the past, not because they were unimportant, but because our minds can only hold so much. And as time moves on, some memories simply slip away, making room for new ones.

But memory is not just about what we recall—it's about how we recall. Often, our memories are selective, capturing the parts that felt significant or relevant at the time. When we look back, we may find ourselves focusing on specific details, revisiting the same moments over and over, while other parts remain hazy. This selectiveness can be comforting, allowing us to hold onto the parts of the past that bring us joy. But it can also be limiting, trapping us in a version of the past that might not be entirely accurate.

What's remarkable is how time itself alters memory. A single day, filled with insignificant details, can, over time, become a pivotal moment in our personal narrative. We look back and assign meaning to events, casting them as turning points or milestones. Yet, at the time, those moments may have felt ordinary. Memory gives us a way to shape our story, to find connections and patterns, to create a sense of coherence in a world that often feels random and chaotic.

There's a particular kind of beauty in the way memories fade. The edges blur, the colours soften, and what remains is a distilled version, a more enduring echo of the original experience. It's as if time has a way of sanding down the rough edges, leaving only what is essential. But this fading is also a reminder of how fleeting our lives are, of how even the most significant moments will one day dissolve into the haze of time.

Nostalgia plays a powerful role in this process. We often look back with a sense of longing, remembering times that felt simpler, happier, or more vibrant. Nostalgia is seductive; it presents a version of the past that is often idealised, casting our memories in a golden light. We remember the warmth of summer days, the laughter shared with friends, the feeling of endless possibilities. But nostalgia is also a form of illusion. It paints a picture of the past that is more about how we want to remember than how things actually were. We gloss over the struggles, the uncertainties, the parts that didn't quite fit. Nostalgia gives us a comforting version of the past, but it's not always the truth.

There are also those memories we'd rather forget, the moments that haunt us, that cling to us despite our best efforts to let them go. Regret, guilt, and pain have a way of embedding themselves in our minds, resurfacing when we least expect them. These memories can be like weights, holding us back, clouding our view of the present. But even these darker memories have their place. They shape us, reminding us of where we've been, of the lessons we've learned, of the mistakes we don't want to repeat.

For all its flaws and imperfections, memory is essential to our sense of identity. Without it, we would be unanchored, adrift in a world where each moment stands alone. Memory connects the dots, creating a narrative that helps us make sense of who we are. It allows us to trace the arc of our lives, to see the paths we've taken, and to understand how we've changed. In this way, memory is time's gift to us—a way to preserve the past, even as it fades away.

But memory is also a reminder of time's relentless march. We can cherish the moments we remember, but we cannot stop them from slipping away. They will fade, just as time moves on, leaving us with the knowledge that life is always in flux. Perhaps this is why we cling so tightly to our memories. They give us a sense of continuity, a way to hold onto something permanent in a world that is constantly changing. Yet, we must also learn to let go, to accept that some things will be lost, that the past is both preserved and fading, just as we ourselves are.

In the end, memory is both a comfort and a challenge. It allows us to look back, to find meaning, to hold onto the moments that matter. But it also forces us to confront the reality of time, to recognise that even our most cherished memories will one day fade. And perhaps, in that fading, there is a lesson—one that reminds us to live fully in the present, to create memories worth keeping, even as we accept that they too will one day be gone.

Chapter 7:
The Anxiety of Time: Confronting Mortality

Time has a way of forcing us to confront our own mortality. From the moment we become aware of its passing, we are reminded that our time here is finite. Each tick of the clock is a step closer to an end we cannot avoid. For some, this realisation is a source of fear, a constant reminder of life's impermanence. For others, it serves as a motivator, pushing them to make the most of every moment. But regardless of how we respond, the awareness of our own mortality is something we all must face.

The fear of death is one of the most deeply rooted anxieties we have. It's a fear that often lurks in the background, influencing our choices, our beliefs, and even our relationships, whether we acknowledge it or not. We may not think about death every day, but it shapes us in ways we often don't realise. It's there in the urge to leave a legacy, to create something that will outlast us. It's there in the desire for control, the need to make our mark before our time runs out. And it's there in the way we cling to life, holding on to the people, places, and things that give us a sense of permanence in an impermanent world.

Yet, for all the fear it inspires, the awareness of death can also be a powerful catalyst for living. When we acknowledge that our time is limited, it can sharpen our focus, helping us to see what truly matters. It can inspire us to pursue our passions, to spend time with those we love, to live with purpose. This is the paradox of mortality: while it may terrify us, it also gives our lives meaning. Without an end, there would be no urgency, no drive to make each moment count.

But living with the awareness of death is not easy. It requires a certain courage, a willingness to face the unknown. Many people cope by pushing it aside, by burying themselves in distractions, by filling their lives with noise and activity so they don't have to think about the silence that waits at the end. Yet, no matter how much we try to ignore it, the anxiety of time will always find its way back to us. It's a shadow we cannot escape, a reminder that life is both precious and precarious.

In the face of mortality, we often find ourselves searching for meaning, for something that will make the inevitability of death more bearable. Some turn to religion or spirituality, finding comfort in the idea of an afterlife, a continuation beyond this world. Others seek meaning in legacy, in the idea that they can live on through their work, their children, their contributions to the world. And still, others find solace in the simple act of living fully, of embracing each moment without worrying about what comes next.

There is no right or wrong way to confront mortality. It's a deeply personal journey, one that each of us must navigate in our own way. Some find peace in acceptance, choosing to embrace death as a natural part of life. They live with a sense of equanimity, knowing that death is inevitable and focusing instead on how they can make the most of the time they have. Others wrestle with it, struggling to come to terms with the idea of their own end, searching for answers in a world that often offers none.

In our culture, death is often seen as a taboo subject, something to be avoided or hidden away. We sanitise it, gloss over it, pretending that it's something that happens only in the distant future, to someone else. But death is not something we can outrun. It's a part of life, as natural as birth, as essential as breath. By refusing to acknowledge it, we do ourselves a disservice. We deny ourselves the chance to make peace with it, to find our own way of understanding it.

The anxiety of time, the fear of death, can feel overwhelming. But it's also a reminder that our time here is a gift, that each moment is an opportunity to live in a way that is true to ourselves. It pushes us to ask the big questions, to consider what we want our lives to mean, to think about how we can leave this world a little better than we found it. Mortality may be inevitable, but it doesn't have to define us. We are more than our fears, more than our anxieties. We are the choices we make, the love we share, the impact we leave behind.

Ultimately, confronting mortality is not about resigning ourselves to death, but about embracing life. It's about recognising that while we cannot control the length of our days, we can control how we spend them. It's about finding the courage to live with intention, to pursue our dreams, to connect deeply with others, to leave behind something meaningful. It's about choosing to live, not in fear of the end, but in celebration of the journey.

In the end, time will take us all. But if we live fully, if we make each moment count, then perhaps we can face that final moment not with regret, but with gratitude. Gratitude for the time we were given, for the experiences we had, for the people we loved, and for the life we lived. Time may be relentless, but it is also a gift, one that we can choose to embrace, even as we acknowledge its limits.

Chapter 8:
Time as Currency: The Economics of Minutes and Hours

Time, much like money, is a currency. It's a resource we spend, save, invest, and sometimes waste. But unlike money, time is finite; once spent, it can never be regained. This makes it both precious and precarious—a constant reminder that every choice carries a cost. We exchange our time for experiences, for achievements, for relationships, and for security. How we choose to spend our time reveals our priorities, our values, and ultimately, our understanding of what life is worth.

In the modern world, time and money are often closely intertwined. We work for hours to earn money, trading our minutes for financial security and material comforts. In this exchange, we are taught to value efficiency, productivity, and output. The more time we invest in our work, the more money we earn, and in theory, the greater our satisfaction should be. But this relationship is not as straightforward as it seems. For many, the pursuit of wealth leads to a scarcity of time, creating a paradox where we spend the bulk of our lives accumulating resources, only to find ourselves with little time left to enjoy them.

This brings us to a fundamental question: how do we value time? In an economic sense, time is often measured in terms of opportunity cost. Each hour we spend on one activity is an hour we cannot spend on another. When we choose to work late, we sacrifice time with family or friends. When we spend our weekends in pursuit of leisure, we forgo potential earnings. These trade-offs are a part of life, but they highlight the underlying tension between time and money. We are constantly balancing the desire for financial security with the need for personal fulfilment, often finding that the two are not easily reconciled.

The value of time also varies depending on our circumstances. For those with financial abundance, time may become the most precious resource. No longer driven by the need to earn, they may choose to spend their time on experiences that enrich their lives, pursuing passions, travelling, or engaging in creative endeavours. For those struggling to make ends meet, however, time can feel like a trap, an unyielding force that pushes them into a cycle of work that leaves little room for anything else. This disparity raises important questions about the economics of time and the ways in which societal structures influence how we spend it.

But beyond these economic considerations, time is deeply personal. Each of us has an inner sense of how we want to spend our days, of what feels meaningful and fulfilling. Yet, too often, this inner sense is drowned out by external pressures. We are told to be productive, to hustle, to maximise every minute. This creates a culture of busyness, where worth is measured by how much we can accomplish in a day. But busyness is not the same as purpose, and a full schedule does not guarantee a fulfilled life.

The true value of time lies not in how much we can do, but in how we choose to live.

To view time as currency is to understand that every moment holds potential. Time is a blank slate, a resource that can be spent in countless ways. We can choose to invest it in relationships, building connections with those we care about. We can spend it on personal growth, pursuing knowledge, skills, and self-awareness. We can use it to create, to bring something new into the world that wasn't there before. Each choice is an investment, a way of using our limited time to create a life that feels meaningful.

Yet, the way we spend our time is not always within our control. Many people find themselves bound by responsibilities that dictate how they use their hours. Work, family obligations, and societal expectations can limit our freedom, leaving us with little choice over how we spend our days. This can create a sense of frustration, a feeling that time is slipping away, lost to activities that don't align with our true desires. But even within these constraints, there is room for choice. We can find small moments to invest in ourselves, to pursue our passions, to nurture our relationships. Time is not just about the grand gestures; it's also about the small moments that, when added together, shape the fabric of our lives.

One of the most powerful ways to reclaim our time is to become aware of how we're spending it. This requires a kind of mindfulness, a willingness to look at our habits and routines and ask whether they truly serve us. Are we investing our time in things that bring us joy, that align with our values, that move us closer to our goals? Or are we spending it on activities that drain us, that fill our days but leave us feeling empty? To see time as currency is to take responsibility for how we spend it, to recognise that each choice is a reflection of what we value.

In this sense, time is both a gift and a responsibility. It's a resource we are given, but it's also something we must manage wisely. How we spend our time determines the quality of our lives, the depth of our relationships, and the legacy we leave behind. To treat time as currency is to understand that it is both finite and precious, that each moment carries a cost, and that once spent, it can never be recovered.

Ultimately, the value of time is not something that can be measured in hours or pounds. It's something that is felt, that is known in the heart. It's the way a sunset makes us pause, the way a conversation with a friend brings us to laughter, the way a quiet moment of reflection reminds us of who we are. Time is the currency of life, the resource that shapes our days and defines our journey. To spend it well is to live with intention, to choose our moments carefully, and to recognise that while we cannot control the quantity of time we are given, we can control the quality of the life we build with it.

Chapter 9:
Suspended Moments: When Time Stands Still

We often speak of moments that seem to defy time, where the usual flow of minutes and hours feels as though it has paused. These are the suspended moments—those rare instances that feel timeless, as if the world has stopped turning just for a second. In these moments, we are completely present, fully immersed in the experience, unbound by the relentless ticking of the clock. Time, for once, feels like it stands still.

These moments can happen anywhere, at any time, and they are usually unexpected. They can be found in the simplest things: the quiet peace of a sunrise, the weightlessness of laughter with someone we love, the awe of standing before something larger than ourselves. Often, they arise when we least anticipate them, as if they are gifts hidden in the fabric of our daily lives, waiting to be discovered. And when they arrive, they have a way of holding us, of drawing us out of the rush and into something deeper.

Suspended moments often come with an intensity that makes them unforgettable. The world around us fades, and all that remains is the here and now, vivid and alive. Time, in these instances, loses its grip, allowing us to feel something beyond its usual boundaries. It might be a moment of pure joy, a pang of sadness, or a rush of gratitude. Whatever the emotion, it feels magnified, as if we are seeing life with a clarity that is usually just out of reach. In these moments, we are not merely passing through time; we are inhabiting it fully.

Some say that these moments are the closest we come to experiencing eternity. There is a quality to them that feels beyond time, as though we've glimpsed something that transcends our usual experience of reality. This is why suspended moments often remain etched in our memories, long after they have passed. They leave a lasting impression, a kind of echo that resonates within us, reminding us of the depth that life holds, even in its most ordinary moments.

Yet, for all their beauty, these moments are elusive. We cannot force them to happen; they arrive unbidden, surprising us with their sudden presence. We might try to recreate them, to go back to the places or situations where we once felt that sense of timelessness, but it's never quite the same. Suspended moments are unique, tied to a particular time and place, a confluence of factors that cannot be duplicated. They remind us of the transient nature of life, of how fleeting and precious each experience truly is.

This is perhaps why we cherish these moments so deeply. They offer us a break from the usual demands of time, a glimpse into a reality where we are not ruled by the clock. In these moments, we feel free—free from the worries of the past, free from the anxieties of the future, free to simply be. It's as if time has stepped aside, allowing us to connect with something essential, something timeless. And in that connection, we find a kind of peace, a sense of belonging, a reminder of what it means to be truly alive.

Suspended moments also have a way of bringing us closer to others. When time stands still, we often find ourselves more open, more present, more attuned to the people around us. Whether we're sharing a laugh with a friend, holding hands with a loved one, or simply sitting in silence with someone we care about, these moments create a sense of connection that goes beyond words. They remind us that time is not just something we experience alone, but something we share with others. And in that sharing, we find a kind of intimacy that binds us together, even if just for a fleeting instant.

But perhaps the most remarkable thing about suspended moments is the way they reveal the richness of life. They show us that time is not merely a sequence of events, but a tapestry of experiences, each one woven with emotion, meaning, and memory. These moments invite us to step back, to see the world with fresh eyes, to appreciate the beauty that is all around us, waiting to be noticed. They remind us that life is more than just a series of tasks and goals; it is a journey filled with moments of wonder, if only we are willing to pause and take them in.

In the end, suspended moments are a reminder of the mystery of time. They show us that, for all our attempts to measure and control it, time is not something we can fully grasp. It moves in ways we cannot predict, sometimes fast, sometimes slow, sometimes not at all. And in those rare instances when time seems to stand still, we are given a gift—a chance to experience life beyond the ordinary, to touch something that feels infinite, to know, if only for a moment, what it means to be truly free.

Perhaps this is the real power of suspended moments: they allow us to step outside of time, to glimpse a world where we are not bound by the usual constraints. They give us a taste of eternity, a reminder that there is more to life than the ticking of the clock. And in that taste, we find a kind of solace, a sense of peace that lingers, long after the moment has passed. Suspended moments may be fleeting, but their impact endures, leaving us forever changed, forever grateful, forever aware of the timelessness that lives within us all.

Chapter 10:
The Elasticity of Time: Why Time Feels Different for Everyone

Time may tick away at a steady pace, yet it doesn't feel the same for everyone. For some, a day can stretch out endlessly, while for others, it can vanish in the blink of an eye. This elasticity of time—its ability to feel long and short, fast and slow—is one of its most mysterious qualities. Although clocks measure time with precision, our experience of it is far from uniform. Time can expand and contract, changing its shape based on how we feel, what we're doing, and even who we're with.

One of the most common experiences of time's elasticity is the way it seems to slow down during moments of intense emotion. In times of danger or excitement, seconds can feel like minutes. Our senses sharpen, our thoughts race, and we become acutely aware of every detail. It's as if time itself has stretched to allow us to process what's happening. This phenomenon is known as time dilation, and while it's typically associated with the world of physics, it also applies to our subjective experience. When we're fully present and engaged, time often feels slower, as if we've been granted a little extra space to take it all in.

Conversely, time seems to speed up when we're caught up in routines or engrossed in something we enjoy. A day at work can feel endless, with each hour dragging along, while a weekend can pass in a flash. This perception is influenced by a number of factors, including our level of engagement, the novelty of our activities, and our state of mind. When we're busy, our attention is fragmented, and we're often not fully present. As a result, time feels quicker, slipping away before we've even noticed. In contrast, when we're doing something familiar or repetitive, time slows, each moment blending into the next.

Psychologists have long studied the ways in which our perception of time changes throughout our lives. For children, time seems to move slowly. Days are long, summers feel endless, and each experience feels new and vivid. As we grow older, however, time seems to accelerate. Weeks blend into months, months into years, and before we know it, entire seasons have passed. This phenomenon is often attributed to the way our brains process new information. When we're young, everything is novel, and our brains are busy encoding each detail, making time feel fuller and more expansive. But as we age, our experiences become more routine, and our brains encode fewer details, creating the sensation that time is speeding up.

Our emotional state also has a profound impact on how we perceive time. When we're anxious or stressed, time can feel painfully slow. Waiting for news, sitting through a meeting, or enduring a difficult conversation can make seconds stretch into what feels like hours. This is because our brains are hyper-focused on the situation, amplifying each moment and making it feel longer. On the other hand, when we're relaxed and content, time seems to flow more smoothly. We're less focused on the clock, more absorbed in our surroundings, and as a result, time feels lighter, more fluid.

The elasticity of time is also influenced by our relationships. Time can fly when we're with someone we love, hours passing as if they were minutes. This is often attributed to the sense of connection we feel, the way we lose ourselves in the presence of another. But time can also feel painfully slow when we're separated from those we care about. A week away from a loved one can feel like an eternity, while time spent together feels fleeting. This contrast highlights the deeply personal nature of time, how it bends and stretches in response to our emotional landscape.

Perhaps one of the most curious aspects of time's elasticity is the way it seems to blur when we look back. A year can feel both long and short, filled with countless memories and yet passing in what feels like a heartbeat. This duality is part of what makes time so difficult to grasp. It is both a constant and a variable, both a measure of duration and a reflection of our experiences. When we reflect on the past, we often find that certain periods feel longer or shorter than they actually were. This is because our brains are not just recalling events but also reconstructing our experiences, adding layers of emotion, meaning, and perspective.

There are also cultural differences in how time is perceived. In some cultures, time is seen as more fluid, less rigidly structured by schedules and deadlines. People are more inclined to live in the moment, to let time unfold naturally rather than trying to control it. In other cultures, time is tightly regulated, with a focus on punctuality, efficiency, and productivity. These cultural differences can shape our perception of time, influencing not only how we experience it but also how we relate to it. For some, time feels like a friend, a gentle guide that moves at its own pace. For others, it feels like an adversary, something to be managed and conquered.

In the end, time's elasticity reminds us that our experience of it is deeply subjective. While clocks and calendars provide a framework, they cannot capture the full complexity of how we live and feel time. Our perception of time is shaped by a myriad of factors, from our emotions and relationships to our culture and stage of life. And while we may never fully understand why time feels different for everyone, we can learn to appreciate its fluid nature, to embrace the way it stretches and contracts, and to find meaning in the moments it gives us.

Understanding the elasticity of time can also help us navigate our lives more mindfully. When we're aware of how our emotions, experiences, and relationships influence our perception of time, we can make choices that allow us to experience it more fully. We can seek out new experiences, cultivate meaningful connections, and take the time to savour the present. By recognising that time is not a fixed entity but a malleable one, we can begin to let go of the need to control it, and instead, learn to flow with it, to embrace its ebbs and flows, and to find beauty in the way it feels uniquely our own.

Ultimately, time's elasticity is a reminder that we are not merely passengers on a one-way journey. We are active participants in the shaping of our time, capable of bending it to our perceptions and experiences. Time may be a constant, but it is also a canvas, one that reflects the fullness of our lives, the depth of our emotions, and the richness of our relationships. To live with an awareness of time's elasticity is to live with an appreciation of its complexity, to embrace the mystery of its passing, and to find peace in the knowledge that while time may change, our experience of it remains uniquely ours.

Chapter 11:
Out of Sync: The Disruption of Time in Modern Life

Modern life has given us the tools to control time like never before, yet we often feel more out of sync with it than ever. We have clocks that measure seconds with precision, calendars that stretch endlessly into the future, and devices that keep us connected around the clock. Despite these tools, or perhaps because of them, many of us feel like we're constantly racing against time, struggling to keep up with its pace. The very fabric of our daily lives seems to pull us away from a natural rhythm, leaving us fragmented, rushed, and perpetually behind.

Technology, which promised to save us time, has ironically become one of the greatest disruptors of our sense of time. In the past, our days were largely governed by natural cycles—the rising and setting of the sun, the changing of the seasons. Now, technology allows us to extend our days far beyond daylight, blurring the line between work and rest. The modern world runs on a 24-hour clock, and with the advent of smartphones and the internet, we are rarely disconnected. The consequence is that we no longer have clear boundaries between work, leisure, and sleep. Time becomes a continuous stream, with fewer and fewer moments of true pause.

This constant connectivity has created an "always-on" culture, where the expectation is that we are available at any moment, ready to respond, react, and engage. Work emails arrive late at night, social media notifications interrupt our meals, and news alerts wake us up in the early hours. The result is a fractured sense of time, where we are always partially somewhere else, divided between the present moment and the digital world that demands our attention. This fragmentation can leave us feeling stretched thin, as if we are living in multiple places at once but never fully present in any of them.

The speed of modern communication also plays a role in this disruption. We are bombarded with information at an unprecedented rate, with news cycles that move so quickly it feels impossible to keep up. Trends change in the blink of an eye, and what was relevant yesterday can feel outdated by tomorrow. This rapid pace can make time feel like it's speeding up, pulling us along before we've had a chance to catch our breath. We are constantly consuming, scrolling, and swiping, racing through the minutes without truly engaging with any of them.

Travel, too, has altered our experience of time. In the past, distance created a natural barrier to how far we could go and how quickly we could move. Now, with the ease of air travel, we can cross continents in a matter of hours, experiencing multiple time zones within a single day. While this has opened up incredible opportunities, it has also disrupted our internal clocks. Jet lag, a term that barely existed a century ago, is now a common experience, a reminder that our bodies are not built to move through time and space so quickly. Even when we're back home, it can take days for our sense of time to realign, leaving us feeling out of sync with the world around us.

The disruption of time in modern life has also affected our perception of productivity. We are constantly told to do more, to be more efficient, to maximise every minute. The rise of multitasking has created an illusion of productivity, where we believe that doing multiple things at once will help us accomplish more. But research shows that multitasking can actually make us less efficient, as our brains struggle to switch between tasks. Instead of saving time, we end up scattered, with our attention divided, our focus weakened, and our sense of accomplishment diminished.

As a result of these pressures, many people find themselves in a state of perpetual hurry, always rushing but never feeling like they're truly getting anywhere. This sense of being out of sync can create a feeling of disconnection, as if we are out of step with ourselves, with others, and with the world around us. We may struggle to find moments of stillness, to connect with the people we care about, or to simply enjoy a quiet moment. The more we try to keep up, the further we seem to fall behind, caught in a cycle that leaves us exhausted and unfulfilled.

Ironically, in a world where time is so carefully measured, many of us have lost our own internal sense of time. We rely on alarms, reminders, and schedules to tell us when to wake up, when to eat, when to work, and when to rest. We have become so accustomed to external cues that we've forgotten how to listen to our own rhythms. The ancient concept of "body time," a natural sense of when to do things based on our internal clocks, has been largely replaced by the artificial structure of the 9-to-5 workday, the weekend, and the holiday. This disconnect can leave us feeling like we are living on borrowed time, always chasing a schedule that doesn't quite fit.

In response to this disruption, some have turned to practices that aim to restore a sense of balance. Mindfulness, meditation, and digital detoxes have become popular as people seek to reclaim their time and reconnect with the present. These practices offer a way to step back from the constant stream of information, to quiet the noise, and to realign with a more natural rhythm. By taking time to pause, to breathe, and to simply be, we can begin to find our way back to a sense of harmony, a feeling of being in sync with ourselves and with the world around us.

Ultimately, the disruption of time in modern life is a challenge that each of us must navigate in our own way. We live in a world that prizes speed, efficiency, and productivity, yet these values often come at the expense of our well-being. To find balance, we must learn to set boundaries, to recognise when enough is enough, and to prioritise what truly matters. We must remember that time is not just something to be managed, but something to be experienced. By slowing down, by being present, by reconnecting with our own internal clocks, we can begin to heal the rift, to find our way back to a more natural, more fulfilling sense of time.

Time, after all, is a gift. It is not something to be conquered or controlled, but something to be cherished. And while the modern world may push us to live out of sync, we have the power to choose our own path, to create our own rhythm, to live in a way that feels authentic and true. We may not be able to stop the disruption of time, but we can learn to navigate it, to find moments of peace amidst the chaos, and to remember that, in the end, time is ours to live as we choose.

Chapter 12:
Borrowed Time: Living with Regret and Anticipation

Time is a gift, yet we often live as though we're borrowing it, spending each moment with our minds tethered to the past or the future. Regret and anticipation are two sides of the same coin, both stemming from our awareness of time's passage. Regret keeps us anchored to what has already been, while anticipation draws us toward what might be. And between these two forces, we find ourselves in a state of borrowed time—living in the present, yet constantly pulled away from it.

Regret is a powerful emotion, a reminder of the choices we've made and the roads we didn't take. It can linger in our minds, casting a shadow over our lives, making it difficult to move forward. We replay moments, wondering how things might have turned out differently if we'd made other choices, if we'd spoken up, taken a risk, or simply been more aware. Regret is a form of borrowed time, a way of living in the past, reliving moments that can never be changed. And yet, regret can also be a teacher, offering us insights into ourselves, showing us what we value, what we fear, and where we still need to grow.

While regret anchors us to the past, anticipation pulls us toward the future. We spend much of our lives looking ahead, imagining what's to come, planning for the days that haven't yet arrived. Anticipation can be a source of excitement, a spark that drives us to dream, to hope, to aspire. But it can also be a source of anxiety, a constant reminder that the future is uncertain, that time is running out, that there are things we want to do but haven't yet done. Living with anticipation can feel like living on borrowed time, as if we're always waiting for something just out of reach, something that will make us whole, something that will finally bring us peace.

Together, regret and anticipation create a tension that defines much of the human experience. We are constantly caught between what was and what might be, between memories that haunt us and dreams that inspire us. This tension can make it difficult to live fully in the present, as we're always looking backward or forward, always borrowing time from what could have been or what could still be. We live in a state of in-betweenness, never fully here, always half somewhere else.

This borrowed time can prevent us from fully engaging with the present moment. When we're consumed by regret, we miss the beauty of what's right in front of us, unable to appreciate the present because we're too focused on the past. When we're lost in anticipation, we overlook the richness of today, seeing it only as a stepping stone to tomorrow. The present becomes a bridge between what has passed and what is yet to come, and in our rush to cross it, we miss the experience of simply being here.

Yet, living with regret and anticipation is not inherently negative. Both have their place in our lives, and both can offer valuable lessons. Regret can teach us to appreciate the choices we make, to be more mindful, to avoid repeating mistakes. It can remind us of the things we care about, the people we love, the dreams we once had. Anticipation, meanwhile, can be a source of motivation, a reminder that there is always more to strive for, always new horizons to explore. It can give us a sense of purpose, a reason to wake up each morning, a vision of what our lives could become.

The challenge is to find a balance, to learn how to live with both regret and anticipation without letting them control us. This means accepting the past for what it is, recognising that it cannot be changed, and forgiving ourselves for the mistakes we've made. It also means embracing the uncertainty of the future, understanding that it is beyond our control, and finding peace in the knowledge that we cannot predict or dictate what will happen. By accepting both the past and the future, we can begin to live more fully in the present, appreciating each moment as it comes, without the need to borrow time from what has been or what is yet to be.

To live in this way requires a certain level of mindfulness, a willingness to let go of the past and the future and to focus instead on the present. This doesn't mean we should ignore our regrets or stop dreaming about the future. Rather, it means learning to live with these feelings without being consumed by them. It means acknowledging our regrets, learning from them, and then letting them go. It means dreaming about the future, but not letting those dreams prevent us from appreciating what we have now.

In the end, borrowed time is a reminder that our lives are finite, that each moment is precious, and that we cannot afford to waste our time dwelling on what might have been or what might still be. We are here now, in this moment, with all its imperfections and possibilities. By embracing this reality, we can find a sense of peace, a sense of gratitude, a sense of presence. We can learn to live with both our regrets and our dreams, knowing that they are part of what makes us human, part of what gives our lives meaning.

Time will continue to move forward, and we will continue to feel its pull. But by learning to live with borrowed time, we can find a way to live more fully, to appreciate the beauty of each moment, to make the most of the time we have. We can learn to be here, now, in this moment, living not in the past or the future, but in the present, where life truly happens.

Borrowed time may be an unavoidable part of the human experience, but it doesn't have to define us. By finding balance, by embracing both the past and the future without letting them overshadow the present, we can learn to live more fully, more deeply, more authentically. We can learn to appreciate the moments we have, to make peace with the moments we've lost, and to look forward to the moments still to come. We can learn to live, not as though we're borrowing time, but as though we are finally, truly, here.

Chapter 13:
The Gift of Time: Finding Meaning in the Moments We Have

Time is both a mystery and a gift, a resource we often take for granted yet one that is undeniably precious. It flows steadily, offering us moments that we can choose to fill with meaning or let slip by unnoticed. Every second we spend is a piece of our lives, a fragment of existence that will never return. In a world that is constantly pushing us to move faster, to do more, and to achieve more, it's easy to overlook the profound value of simply being present in the moments we have.

Finding meaning in our moments starts with recognising that time itself is a gift. We are given a finite number of moments, each one an opportunity to experience, to connect, and to grow. The challenge is to learn how to value these moments, to see them not as things to be used up or spent, but as treasures to be appreciated. This shift in perspective can change the way we live, transforming the ordinary into something extraordinary, infusing our days with a sense of purpose and gratitude.

Living with an awareness of time as a gift requires us to slow down and pay attention. It means pausing long enough to notice the details, to feel the texture of life, to appreciate the people and experiences that fill our days. Too often, we rush from one task to the next, so focused on what's ahead that we miss the beauty of what's right in front of us. But when we take the time to be fully present, we begin to see that even the smallest moments can hold immense value. A shared smile, a quiet walk, a moment of stillness—these are the moments that give life its depth and richness.

Finding meaning in our moments also means learning to let go of distractions. In our hyper-connected world, it's easy to get caught up in the constant stream of notifications, updates, and information that demands our attention. We spend so much time looking at screens, checking messages, scrolling through feeds, that we lose touch with the present. By choosing to disconnect, even briefly, we can create space to reconnect with ourselves, to listen to our own thoughts, to engage with the world around us. This act of being fully present allows us to experience time in a more personal, more meaningful way.

One of the most profound ways to find meaning in our moments is to cultivate a sense of gratitude. When we appreciate the time we have, we begin to see life as a series of gifts, each one unique and valuable. Gratitude helps us to focus on what we have rather than what we lack, to recognise the abundance in our lives rather than the gaps. It shifts our perspective, allowing us to see the blessings in even the most ordinary moments. By living with gratitude, we find joy in the simple things, a joy that is not dependent on external achievements but rooted in a deep appreciation for the present.

Another way to find meaning in our moments is to engage deeply with the people around us. Relationships are one of the greatest sources of meaning in our lives, and yet, they are often the first things we neglect when we're busy. Taking the time to truly connect with others, to listen, to share, to support, can transform our experience of time. In these moments of connection, we find a sense of belonging, a reminder that we are part of something larger than ourselves. We find that time spent with loved ones is never wasted, but rather, it is the foundation upon which our most meaningful memories are built.

Living with an awareness of time as a gift also involves embracing the present without fear of the future or regret for the past. It means letting go of the need to control what's coming next and accepting that each moment is enough, just as it is. When we focus too much on the future, we miss the opportunities that are right here, right now. And when we dwell on the past, we lose sight of the potential for joy and growth in the present. By choosing to live fully in the now, we can make the most of the time we have, finding meaning not in what was or what might be, but in what is.

Meaning in the moment is not about grand gestures or life-changing events; it's about finding significance in the everyday. It's about seeing the beauty in a quiet morning, the warmth in a shared laugh, the peace in a deep breath. These are the moments that may seem small, but they are the building blocks of a meaningful life. When we learn to see them for what they are, we realise that life is not something to be chased or conquered, but something to be embraced, moment by moment, with an open heart.

To live with an awareness of the gift of time is to live with intention. It is to recognise that while our time is limited, our ability to create meaning is infinite. We can choose to spend our time on things that align with our values, on activities that bring us joy, on people who lift us up. We can choose to make each day a reflection of what matters most to us, creating a life that is not just lived but felt deeply. In doing so, we honour the gift of time, making the most of the moments we have, and creating a legacy that will outlast us.

In the end, finding meaning in the moments we have is about being fully alive. It's about recognising that life is not measured by the number of breaths we take, but by the number of moments that take our breath away. It's about living with a sense of wonder, a sense of purpose, a sense of connection. Time may be fleeting, but within each moment lies the potential for something beautiful, something profound, something that makes life worth living.

Time, as a gift, invites us to be present, to be grateful, to be engaged. It invites us to find meaning not in what we possess or what we achieve, but in how we choose to live. And in that choice, we find the essence of what it means to be human, to love, to learn, to grow. To recognise the gift of time is to see life in all its complexity, to embrace its fleeting nature, and to cherish each moment as if it were our last. Because, in the end, the moments we have are all we truly own, and it is up to us to make them count.

Chapter 14:
The Timeless Self: How We Outlive Our Own Time

While time shapes much of our experience, there's a part of us that seems to exist beyond its reach—a sense of self that feels timeless. This is the part of us that remains constant amid change, the essence that we carry with us throughout life. It is the part of us that questions, dreams, loves, and connects. This timeless self is not bound by the ticking clock or the passing years; it is an enduring presence, a reminder that while our physical lives are finite, our impact can extend far beyond our years.

The notion of a timeless self invites us to consider what we leave behind. While our bodies age and our circumstances shift, there are aspects of our lives that continue to ripple outward. These legacies are the ways in which we outlive our own time, the ways in which our presence endures even after we're gone. They are the words we speak, the love we share, the ideas we champion, and the influence we have on others. Through our actions, we leave traces of ourselves in the lives we touch, building a legacy that echoes through time.

Legacy is often associated with grand achievements, with names etched in history books or statues erected in honour of remarkable lives. But legacy can be far subtler, far more personal. It can be the wisdom we pass on to our children, the kindness we show to strangers, the ways we make others feel seen and valued. Our legacy is woven into the lives of those around us, carried forward in ways we may never fully understand. Even the smallest acts can leave an imprint, a reminder that our time here was meaningful.

This concept of a timeless self also challenges us to think about the values we uphold, the principles by which we live. What are the things that matter most to us, the things we want to be remembered for? These questions invite us to reflect on the kind of person we want to be, not just for ourselves, but for those who will remember us. They remind us that our choices shape not only our lives but also the lives of those who come after us. In this way, our actions extend beyond our own time, influencing the future in ways we might not anticipate.

One of the most profound ways to connect with our timeless self is through creativity. When we create—whether it's through art, music, writing, or simply building something new—we tap into a part of ourselves that is unbounded by time. Creativity allows us to express our deepest thoughts, our most intimate feelings, our unique perspectives. These creations become a form of legacy, something that lives on after us, something that others can experience, enjoy, and be inspired by. Through creativity, we leave behind a piece of ourselves, a reminder that we were here, that we had something to share.

Relationships also form a crucial part of how we outlive our own time. The connections we forge, the bonds we nurture, and the love we give all contribute to our legacy. In the end, it is often the relationships we build that define us, that carry the essence of who we are into the future. Friends, family, colleagues—all those we interact with—carry a part of us with them. The memories we create together, the support we offer, the laughter we share—these are the things that linger long after we're gone. Our relationships are living legacies, reflections of our timeless selves that continue to grow and evolve even in our absence.

The concept of the timeless self also invites us to consider the impact of our beliefs and ideas. Throughout history, individuals have left lasting legacies through the ideas they championed, the causes they fought for, and the visions they held. These legacies remind us that while time may limit our physical presence, it cannot contain the power of our convictions. When we stand up for something we believe in, when we work toward a cause greater than ourselves, we contribute to a legacy that outlives us. We become part of a larger narrative, a story that continues to unfold long after we've played our part.

Understanding the timeless self also involves coming to terms with the idea of mortality. While our lives are finite, the ripples we create are not. There is a kind of peace that comes with recognising that, while we may not always be here, our presence will endure in the lives we've touched, in the contributions we've made, and in the legacy we leave behind. This awareness can free us from the fear of time, allowing us to focus not on how long we live but on how fully we live. It reminds us that the quality of our time is what truly matters, that we have the power to shape our legacy through the choices we make every day.

Ultimately, the timeless self is a reflection of how we choose to engage with life, how we respond to the world, how we invest our energy, and how we express our unique gifts. It is the part of us that exists beyond the constraints of time, the part that remains long after our physical presence has faded. By connecting with this timeless self, we can live with a greater sense of purpose, a deeper appreciation for the present, and a more profound understanding of our place in the world.

The timeless self is a reminder that, while time may be relentless, we are more than just passengers on its journey. We have the power to create, to influence, and to inspire. We have the ability to outlive our own time, to leave a legacy that transcends the years, to shape a future we may never see. And in doing so, we find a kind of immortality, a way of living beyond the limits of our days, a way of ensuring that we, too, become timeless.

Chapter 15:
Beyond the Clock: Rethinking Time's Hold on Us

For most of us, time is an invisible force that shapes our lives, a constant presence that governs our schedules, our goals, and even our identities. We measure our days by the clock, our years by the calendar, and our achievements by milestones. But what if we could go beyond the clock? What if we could rethink time's hold on us and reclaim it as something less rigid, less confining, and more aligned with our true selves?

The concept of going beyond the clock invites us to reconsider our relationship with time. Clocks and calendars are useful tools, but they are not the full story. They measure time, but they do not define it. Time, in its essence, is not something we can capture or control. It flows naturally, following rhythms that are both familiar and mysterious. By recognising this, we can begin to see time not as a cage that traps us, but as a river that carries us. We can learn to move with its current, rather than fighting against it.

One way to rethink time is to focus on experiences rather than hours. Instead of asking ourselves how much time we have, we can ask how we want to spend it. We can choose to invest our time in experiences that bring us joy, that nourish our relationships, and that help us grow. By shifting our focus from quantity to quality, we free ourselves from the tyranny of the clock and open up space for moments of true connection and fulfilment. We begin to see time not as something we lose, but as something we gain with every meaningful experience.

Another way to go beyond the clock is to embrace the concept of timelessness. There are moments in life—those suspended moments—where time seems to stand still. These are moments of deep presence, where we are fully immersed in what we are doing, without concern for the past or the future. By cultivating mindfulness, we can create more of these moments, allowing ourselves to step outside the constraints of time and into a space where we are simply being. In this space, we find a kind of freedom, a sense of peace that transcends the usual demands of our schedules.

Rethinking time also means letting go of the idea that time is something we must conquer or control. We often view time as an enemy, something that is always running out, something that we must race against. But time is not our adversary. It is a constant companion, a reminder that life is fleeting and that every moment is precious. By accepting time as it is, rather than trying to bend it to our will, we can find a sense of harmony, a way of living that is less about doing and more about being.

In the end, going beyond the clock is about reclaiming our power over how we experience time. It is about choosing to live in a way that is true to ourselves, that honours our values, and that reflects our deepest desires. It is about realising that time is not something that happens to us, but something we actively engage with, something we can shape through our choices, our attitudes, and our actions. By embracing this perspective, we can begin to live more fully, more authentically, and more freely.

Conclusion

Throughout this journey, we've explored time from many angles—its nature, its elasticity, its impact on our lives and our sense of self. We've seen how time can feel like both a friend and an adversary, how it shapes our memories and our dreams, how it challenges us with its brevity and rewards us with its possibilities. We've confronted the anxiety of time, learned to appreciate its gift, and discovered ways to outlive our own moments by leaving a legacy that echoes beyond our years.

Time is, ultimately, both relentless and forgiving. It moves forward without pause, yet within each moment lies an opportunity to pause ourselves, to reflect, to be present. The true power of time lies not in its control over us, but in our ability to find meaning within it. Whether we live in harmony with time, feel at odds with it, or learn to go beyond it, our experience of time is uniquely our own, shaped by how we choose to perceive it and live within it.

As we move forward, we carry with us the understanding that time is both a measure and a mystery. We cannot hold it, but we can cherish it. We cannot slow it, but we can savour it. By embracing time in all its complexity, we open ourselves to the fullness of life, to a way of being that honours both the fleeting and the eternal.

End Note

Time is what we make of it. It is a canvas, a stage, a river. It is the backdrop of our lives, the rhythm of our days, the thread that ties our moments together. We may not control it, but we do have the power to shape our experience of it. So let us live with intention, let us love without restraint, let us make every moment matter. Because in the end, time is not just a measure of our lives; it is the essence of our existence. And within each tick of the clock lies the potential for something extraordinary.

www.ingramcontent.com/pod-product-compliance
Lightning Source LLC
Chambersburg PA
CBHW070319220526
45465CB00004B/1907